Air

written by Maria Gordon
and
illustrated by Mike Gordon

Wayland

Simple Science

Air

Colour

Day and Night

Heat

Electricity and Magnetism

Float and Sink

Light

Materials

Push and Pull

Rocks and Soil

Skeletons and Movement

Sound

Series Editor: Catherine Baxter

Advice given by: Audrey Randall - member of the Science Working Group for the National Curriculum.

This edition published in 1995 by
Wayland (Publishers) Ltd

First published in 1994 by
Wayland (Publishers) Ltd
61 Western Road, Hove
East Sussex, BN3 1JD, England

© Copyright 1994 Wayland (Publishers) Ltd

British Library Cataloguing in Publication Data
Gordon, Maria
Air.- (Simple Science Series)
I. Title II. Gordon, Mike III. Series 551.51

HARDBACK ISBN 0-7502-1290-X
PAPERBACK ISBN 0-7502-1726-X

Typeset by Jonathan Harley
Printed and bound in Italy by G Canale

Contents

Air is all around us. We cannot see it, so how do we know it is there? We can feel and hear it when it moves.

We can see the things it moves.
Sometimes air even makes
us move!

There is air in almost everything around you.
See for yourself...

Stuff a napkin in the bottom of a plastic cup.
Turn the cup upside down. Put it in a sink full
of water. Now slowly lift it out. Do not tip it.
The napkin stays dry! The water cannot go
into the cup because it is full of air.

What happens if you tip the cup? The air is pushed out of the cup to make room for the water. What happens to the napkin?

Even bricks have air in them!
Fill a bowl with water
and put a brick into it.
What comes out
of the brick?

When you see bubbles you are watching air move. Leave water in a glass for an hour or more. Do you see bubbles on the glass? Air is in the water.

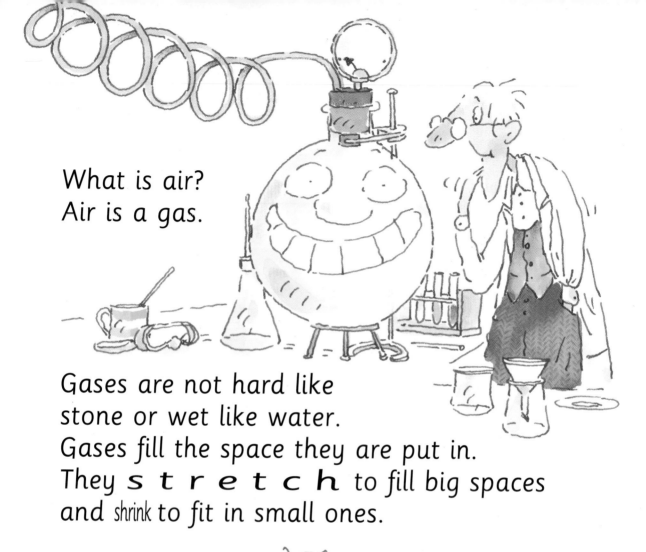

What is air?
Air is a gas.

Gases are not hard like
stone or wet like water.
Gases fill the space they are put in.
They s t r e t c h to fill big spaces
and shrink to fit in small ones.

10

Air is a gas made from lots of gases
mixed together. The main gases
are nitrogen and oxygen.

A long time ago, a Frenchman called Lavoisier found out that the oxygen in air is needed to make things burn. You can see this for yourself.

Stick a candle to a dish with some Plasticine.

Pour water into the dish.

Ask a grown up
to light the candle.

Cover the candle
with a glass jar.

Watch what
happens.

The candle goes out when it has used up all
the oxygen in the jar. This makes room for
the water, so the water rises up in the jar.

We need oxygen, too. We breathe oxygen **in** and we breathe **out** a gas we don't need called carbon dioxide.

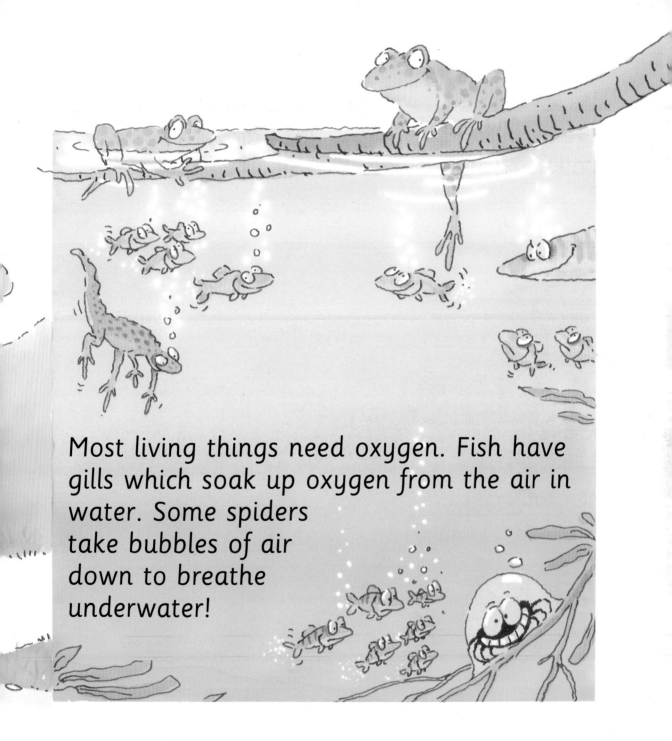

Most living things need oxygen. Fish have gills which soak up oxygen from the air in water. Some spiders take bubbles of air down to breathe underwater!

oxygen

carbon dioxide

It may surprise you to hear that plants breathe too! But, unlike us, they breathe **in** carbon dioxide and breathe **out** oxygen.

Plants breathe through tiny holes in their leaves.

Select four living leaves. Block the air holes by smearing both sides with Vaseline. In a day or two the leaves go pale.

Wipe the Vaseline off two leaves. The clean leaves get better but the other two die.

The air that surrounds
the Earth is called the
atmosphere. The air is
thick close to the Earth.
Higher up, the air
spreads out more
and more until
there is none
left - only space.

The atmosphere stops too much sunlight from getting to Earth, and too much heat from leaving Earth. Without air, our blood would boil in the daytime and freeze at night!

Air is heavy!

Hang a long piece of wood
from its middle. Pump
up two balloons
and tie one to
each end of
the wood.

Burst one balloon and see how
the other one tips the wood.
The air inside makes the
balloon heavier.

When air gets warm it spreads out. This makes it lighter, so it rises. Trace this shape and cut along the dotted lines. Hang it over a heater. The warm rising air makes the shape turn.

Hot air holds up big balloons. Birds ride up on warm air without flapping their wings.

Cold air rushes into the space left by the rising warm air.

Warm and cold moving air
makes winds. Winds carry
our weather! Some carry
clouds and bring rain.
Others push clouds
away and let the
sunshine in.

Push a small piece of paper tissue into the neck of a bottle. Try to blow it in. You can't, because the air in the bottle is pushing back harder than you can blow!

The air inside tyres pushes so hard that they can hold up huge lorries. Air pressing and pushing is called air pressure.

Moving air does not push or press as much as still air. Ask a grown up to cut a piece of paper about 3 cm by 20 cm. Hold one end just under your bottom lip. Blow hard.

What happens? Your paper is like a wing.

Wings push air out of the way.
Air rushes over them and stops
pressing down so much. But the
still air under wings is pushing
up. This keeps wings up. It is
why birds and planes can fly.

Fast air shooting out of a balloon makes the balloon move.

This is how jet planes fly fast.

Moving air can turn machines to help us make things.

Winds carry water, seeds and insects round the world.

Air even carries germs and dirt!

28

The dirt in air goes into buildings, plants and our own bodies. If we don't look after our air and keep it clean, the world will soon be in big trouble! How many different ways can you see air being used here?

Notes for adults

The Simple Science series helps children to reach Key Stage 1: Attainment Targets 1–4 of Science in the National Curriculum. Below are some suggestions to help complement and extend the learning in this book.

4/5   Write a story about the pictures. Make paper fans. Blow into recorders.

6/7   Sit on squashy cushions. Watch smoke trails. Blow bubbles. Whisk air into cake mixtures. Leave boiled water standing – it creates no bubbles.

10/11   Melt some ice and boil away some of the water - three states of matter. Investigate nitrogen in soil, crop rotation and fertilizers.

12/13   Research some ancient and medieval concepts of the four elements: earth, air, fire and water. Look into the use of canaries as gas alerts in mines.

14/15   Borrow an oxygen cylinder and scuba gear. Write a play that is set underwater or in space.

16/17   Discuss pollution. Contrast picture displays of forests and industrial chimney stacks. Use dry ice.

18/19   Pinpoint the highest mountains in the world. Show pictures of climbers with breathing apparatus. Borrow double glazing samples to show air gaps. Wear string vests! Use Thermos flasks. Feel fur and feathers.

20/21 Watch smoke rising and heat shimmers (mirages).

22/23 Write a hot air poem. Borrow a weather map showing the paths of winds over hot and cold parts of the Earth.

24/25 Hold down one end of a ruler with a sheet of newspaper. Press down on the end that is sticking out over the edge of the table.

26/27 Display pictures of windsurfers, parachutists, hang-gliders etc. Make paper planes. Hunt for airborne seeds on a nature trail.

28/29 Visit windfarms/windmills. Discuss and pinpoint ozone holes on a map. Trace bird and butterfly migration paths. Blow noses - they filter dirt. Go to a hot-air balloon show!

Other books to read

Atmosphere by John Baines (Wayland, 1991)

Breathing by Joan Gowenlock (Wayland, 1992)

My Science Book of Air by Neil Ardley (Dorling Kindersley, 1991)

Science Magic with Air (Watts, 1993)

Wind by Joy Palmer (Watts, 1992)

Index